MOTOR CARS 1945 - 1964

CLASSIC MARQUES VOLUME ONE

Alan Earnshaw & Robert Berry

NOSTALGIA ROAD

Is published jointly by
Atlantic Transport Publishers
Trevithick House
West End, Penryn
Cornwall, TR10 8HE
Tel. 01326 373656 Fax. 01326 378309
&
Trans Pennine Publishing
PO Box 10
Appleby-in-Westmorland
Cumbria, CA16 6FA
Tel. 017683 51053 Fax. 017683 53558

Layout & Design
Kevin Bradley
Burwain Studios, Penrith

Reprographics
Barnabus Design & Repro
Threemilestone
Cornwall, TR4 9AN

And Printed by
The Amadeus Press
Leeds Rd, Huddersfield

CONTENTS

INTRODUCTION 2

CHAPTER 1
The Upright Fords 8

CHAPTER 2
The Ford Pilot V8 11

CHAPTER 3
Consul, Zephyr, Zodiac Mk I 14

CHAPTER 4
The 100 E Series 19

CHAPTER 5
Consul, Zephyr, Zodiac Mk II 22

CHAPTER 6
The Anglia 105 E / 123 E 27

CHAPTER 7
The Prefect 107 E 31

CHAPTER 8
Consul Classic & Classic Capri 32

CHAPTER 9
Zephyr, Zodiac Mk III 37

CHAPTER 10
Cortina Mk I 41

CHAPTER 11
Corsair 45

ISBN 0-906899-81-8
British Cataloguing in Publication Data
A catalogue record for this book is available from the British Library

NOSTALGIA ROAD

FORD MOTOR CARS 1945 - 1964

INTRODUCTION

One surprising factor of recent times is the growth of the nostalgia market, a backward glimpse not into history but to our own recent past. The subject can be as diverse as old cigarette cards to enamelled advertisements, or from items as small as comics to things as large as steam locomotives - a subject more commonly dealt with in my writings. However, to those of us who belong to the post-war generation, perhaps the most significant singular influence in our lives has been the automobile. Whether our parents had a car, or even if we are motorists or not is rather immaterial, for the mass car-ownership which began as a trickle in the 1950s and turned into the deluge of the '60s has by now affected all of us. Whether this be through the demise of our public transport systems, or through global warming, the car has changed our lives in no uncertain terms.

Front Cover: *A 1956 Ford Zodiac 206E produced in a limited colour scheme, and pictured here in period setting. This picture was used in Ford's official publicity material of the day.*
Rear Cover Top: *Another Mk II Ford from 1956, this time a Zephyr model in what appears to be Hereford Green livery. The picture also demonstrates the spacious bench seats.*
Rear Cover Bottom: *One of the prestigious models from 1965 was the Corsair, pictured here in yet another publicity photograph.*
Title Page: *Where the majority of British Ford's began the production line at Dagenham; in this view a batch of Prefect E93A saloons roll down the line, although at least one light commercial can be found in the picture.*
Picture Left: *Ford's original British works at Trafford Park, Manchester with a variety of Model T's outside the works.*
Portrait This Page: *Henry Ford the man who introduced motoring for the masses: 30th July 1863 - 7th April 1947.*

The growth of car ownership, the freedom of greater mobility and the social revolution that followed the Second World War has shaped the way of our society today. The growth of express coach services and the contracting size of the rail network and rural bus services being the result. The advent of car ownership has geared society to live at a much faster pace, it has also provided the criminal element of our race with a greater means of escaping detection and punishment. Whatever the case, the car is here to stay. Of all the major manufacturers, the Ford Motor Company has done the most to introduce motoring for every one, the pages that follow tell their story in the period from 1945 to 1965.

The First Ford Car Known As The Cycle Car And Produced In1896

Car ownership is common as we near the end of the 20th century, most households have at least one, but more frequently two and three cars per house are becoming the norm. Yet it is not that long ago that car-ownership was a luxury, and within living memory many of us recall that the only way to get around was by bus, train or on foot. Yet, in the period following World War II, cars became far more readily obtainable, as the 1940s and '50s progressed, cars (like television) became the thing that most families desired to possess. The achievement of car-ownership was, in no small way, due to the progressive policies of the larger manufacturers, and here in Britain firms like Ford, Austin and Morris, did more than their share of making this dream of eutopia come to fulfilment. In the following chapters we look at how Ford helped the 'every-day man' achieve his dream, by producing models that were both economical to buy, and reliable in performance. In the post-war years they built on the tradition of mass-production and keen pricing, that they had achieved in the earlier decades and by models pictured in this chapter.

For the majority of people the car meant ease in getting to and from work, less expensive holidays and day trips into the country or to the seaside. It also created the unknown phenomena of traffic jams, particularly on Bank Holiday weekends as the industrial and commercial centres of Britain teemed out their masses on the seaside resorts like Brighton, Blackpool, Southend and Scarborough. Narrow winding roads in the Yorkshire Dales, the Lake District and the West Country were choked, and the market towns became bottle-necks which often took more than an hour or so to pass through. Places like York and Malton on the A64, Taunton and Tavistock on the road west and Crawley and Reigate on the A23 came to be major bottle necks, but perhaps the worst of the lot was the A6 with infamous delays at Buxton, Lancaster and Preston.

Model A 1903

So it came to be, that the words by-pass and motorway first began to be widely used in the English language - every major town it seemed wanted a by-pass, so contradictory to the earlier trends of taking all the roads through the town centre to capture trade. The first of these was the Preston by-pass which opened in 1959, and became Britain's original section of motorway, now part of the M6.

Those who were drivers or passengers in those days will no doubt recall the great traffic jams that bedevilled our road network in pre-motorway days. For some it was the towns of the Great North Road, or Penrith on the A6, with its notoriously narrow section between Devonshire Street and Middlegate. For others it was the interminable slog down to the west country, particularly between Bristol and Exeter, which occasioned conditions where frustrated fathers would have to quieten fractious children. Games like 'Count the buses', 'I-Spy', and 'See how many different types of car you can see' became part of the way of life to

Model B 1905

Model C 1905

Model K 1907

children of the '50's and early '60's. The latter was usually the favourite, for in those days there was offered an infinite variety of British-made cars, each with its own distinctive style, badge, and nickname. Of these, perhaps the 'four wheels and a board' of the Ford Motor Company were amongst the most common, with the 'sit-up and beg' styling and uninspirational colour schemes of the post-war period. In our games I usually chose Fords, because in that way I was usually sure to win; one of the company's models could normally be seen at every twist and turn! The preponderance of Ford cars was probably due to the fact that the company had prolonged the original concept of its American founder, Henry Ford, who is pretty famous for two sayings - 'Motoring for the masses', and 'You can have any colour you like, just as long as it is black'. In the accompanying photographs and the paragraphs that follow we will show how Ford developed its product range and achieved its low-cost motoring models. The selection of small photographs in this chapter showed how they helped to bring motoring within the reach of the 'ordinary' man here in Britain.

Model R 1907

Model T 1909

At the outset it is perhaps better to qualify the authors' specific interests in the subject, for both of us enjoyed a childhood in which Ford cars figured largely. In Robert's case the family car was one of the old up-right models, a Prefect E493A (JDK 807). Even years after its roving days were done, the old car sat in a corner of a field on the family farm - providing a shelter for the hens, and an adventure playground for the Berry children. His childhood is also filled with memories of his uncle's Squire estate car, which ran beautifully for several years and took them on many excursions. Our initial family car was an Austin A35, but our 'Ford experience' first came with an Anglia 105E Super, registered number 1604 NA which finally arrived in August 1961 after we had endured the infamous waiting lists for 11 months. The Anglia lasted two years and was eventually replaced with a Consul Classic (76 VMB), of 1962 vintage. Though second-hand, this was one of the batch with which Ford suffered so many teething problems. In fact, when my father bought it as a year old car for £300 it had just 121 miles on the clock - as well as a replacement 1500cc engine which was barely run in. It came to a sticky end when it met head-on with a National Coal Board wagon of Foden origin on a foggy September morning in 1966; fortunately this altercation took place directly outside the front entrance of Doncaster Royal Infirmary. Our replacement was UVH 21 a 1963 model of the Classic, though converted as a Lotus Classic and not actually put onto the road until January 1965.

Model T 1911

Model T 1912

Model T 1914

This rare car came to us with just 1,239 miles on the clock, having only ever seen the interior of various northern Ford showrooms and though it had been entered in one race - it too came to a grievous end, one from which my fiance (now my wife) and I were lucky to escape unscathed when it suffered a simultaneous rear-tyre blow-out opn the way back from the Wembly Assembly in July 1969. After that my allegiance turned once again to Austin/British Leyland/Rover cars, and has remained so down to today. Robert on the other hand has spent the several years being involved in the restoration of two Mk.II Consuls, one a Highline Model and one a Lowline Model.

Model T 1917

This book is perhaps our own personal trip back down Nostalgia Road and into our childhood; days when cars had character and individuality. Days when the RAC or AA patrolmen respectively saluted you from the roadside, providing you were displaying your gleaming chrome badge on the bumper or radiator. There were no relay-vehicles, no home-start schemes, and not a great deal of reliability, but above all those days were fun. They weren't perhaps the pioneering days of motoring, but they were the days when cars were available to people whatever their rank and station - upper class, middle class, working class - there was a car for everyone. For as well as the new car showroom agencies, this was also the era of the back-street used car lot. In such back-street establishments you could be sure of finding almost anything, ranging from the mundane pre-war family saloons to exotic American saloons or continental cars which varied in their style and quality.

Model T 1923 Fordor Saloon

This post-war era was also the time when you could buy a used car and drive away in sure confidence, the sure confidence that the guarantee of some salesmen meant nothing! Most were decent businessmen, but others were frankly little more than spivs. Alarmed at the proceedings of some of these back-street traders, the motor manufacturers and the reputable dealers established customer care programmes and training schools for the people employed in promoting, selling and repairing their models. This was an area where Ford of Britain took an important lead!

Model T 1927

Model A Tourer 1932

For many of the working class this was often a period when they made that transition from the bus queue or railway platform to the Triumph, Velocette or BSA motorcycle combination. From these they often went on to putting down their deposit for an Austin A35 or a 'Sit-Up-And-Beg' Ford Popular! It was a time for the home mechanic, and the family washing the Anglia in the drive ready for the weekend's outing.

At the other end of the scale, there was the supreme elegance of a Rolls Royce whilst some, like the ostentatious and flamboyant Lord and Lady Docker with their customised Daimlers, went to extreme lengths of fashion. However, to the man in the street, these cars seemed a world away and, as we will see later, Ford were ready to provide an affordable alternative. It was also an era where motor-sports became even more popular, and few could fail to be thrilled by the exploits of Messrs Hawthorne, Moss, Collins and other schoolboy hero's at the Silverstone race-track. The Monte-Carlo Rally, the ultimate saloon car endurance race of the day, was there for almost anyone who wanted to 'Have-a-Go' - provided they had the money of course!

Model A Saloon 1932

Model F Tourer 1932

Few European cars were on Britain's roads and, of those that were there, most were hardly ever noticed. They were not the commonplace sight they are today! However, a few models were just starting to make in-roads, especially the Renault Dauphine from France and the Volkswagen Beetle from Germany, whilst the land of gondoliers and ice-cream men offered us the Fiat!!!

It was a time when every small boy began to know his cars, either from his I-Spy books or from the Dinky Car catalogue. Other children learned to be wary of the increased traffic on the road, especially when walking slowly back from an ice-cream van! By the 1960s, these vehicles were often Ford Thames vans converted especially for the purpose and bearing sign-writing drawn with a flourish that pronounced a name which often ended with an 'i' (why didn't they buy Fiats we wonder). As raspberry sauce dripped onto little fingers, policemen tried to get across the message 'Mind That Child'. As these officers stood sentinel outside parks, schools and on street corners, their cars would be parked at the roadside. Their new patrol cars invariably would, without doubt, have been black and British; Ford, Riley, Wolseley and Austin being amongst the most common makes employed by the guardians of the law, who were rapidly making the change from push bikes or flat feet to the luxury and speed of fast motor vehicles.

Model B 1932

Model Y Popular 1933

Throughout the two decades which followed the Second World War, the automobile influenced our lives. It entered our fantasy lives too, a fact that almost certainly came about by the growing media of the television screen. Interestingly, thirty years on, nearly everyone remembers the big Fords in Z-Cars, or the lovely Emma Peel with her Lotus Elan. In the same programme, The Avengers, John Steed's vintage Bentley became as much a star as the actors, meanwhile the Saint's Volvo brought new devotees to the Scandinavian motor manufacturer. They were as popular in their day as Lovejoy's Morris Minor and the Jaguar Mk.II driven by John Thaw as Inspector Morse in more modern programs.

V8 30HP 1934

V8 30HP 1936

The period covered by this book was not only a time of change within our own lives, but a growing up and a time of expansion. It was also, sadly, sometimes a death within the British motor manufacturing industry as names like Alvis, Jowett, Armstrong Siddeley and a long list of great firms closed their door for the last time. Fortunately, examples of these Marques have survived and are still to be found with caring owners. The Ford Motor Company has, in a way, been a family of cars with what was (during the period covered by this book) often quite a subtle change from one model to another, rather than a series of radical developments. Although entering the 1930s with the assembling of mainly American-inspired models, the firm's introduction of the model 7Y (an 8 hp family saloon) showed what Ford could design and build in this country! By 1937, Ford of Britain introduced the 10hp model 7W, after which Dagenham became less dependent on its American parent for its new models. As the pictures on this page show, the models that Ford were producing in the 1930s, had now developed a more stylish look, and were considerably less-angular in appearance than the models produced in the first three decades of this century. Attractive saloons, for both family and business use featured in the V8 models and the 10 horse-power Fords.

Model 62 V8 22HP 1936

Model CX 10HP 1934

Inside a couple of years the company had moved on, and both 7Y and 7W models were replaced in 1939 by the E04A and the E93A models respectively. These new cars were similarly bodied from the windscreen back, and as we will show later both these would (in turn) later evolve, with cosmetic and mechanical improvements, into the E494A and the E493A models of the post-war years. As for the larger cars, what then? Well after the various American inspired V8's, like the Model 62, Model 90, and Model 91 etc., it was a time for change. This led us to the post-war V8 Pilot, and the introduction of new 'British' blood. In turn this resulted in the Consul and Zephyr of 1950, which came as part of the Ford Motor Company's aim to provide continual improvements while offering sound value for money, and a car to suit all income brackets and all needs. Perhaps that was Ford's goal, for they produced almost everything, with only a really top-flight luxury car missing from their range. That range is portrayed in the following pages, covering the company's models from 1945 to 1964 and what was arguably the most successful Ford range ever built, the Cortina; the first version of which is pictured opposite. The changes that were progressively introduced by Ford in these twenty years were, undoubtedly, a major part of the golden years of motoring.

Deluxe 10HP 1937

Ford Eight 1937

In the production of this book the authors would like to thank **Ford of Britain**, *for their kind assistance in making this work possible; with particular appreciation to Simon Sproule, Ron Staughton and Fran Chamberlain, It is also appropriate to thank the many Ford owners, dealers and classic car clubs for their kind assistance, although these are far too numerous to mention by name.*

FOO 836B

Picture Above: *A view of the 'Final Assembley Line' at Ford's Dagenham Factory. This plant was built on the side of the Thames Estuary, on what was little more than reclaimed mud-flats, but it went on to become a massive complex with large areas of associated worker's housing being developed in the surrounding countryside. The view shows the EO4A Anglia's having a final Quality Assurance inspection before being dispatched to the dealerships.*

CHAPTER 1
THE 'UPRIGHT' FORDS

By 1955, the products in Ford's British range were almost of entirely modern designs, the only exception was the 103E Popular. This was a small family saloon that was specifically introduced for the 1953 Earls Court Motor Show! The evolution of the Popular 103E came very much from pre-war designs, but it was aimed at the post-war motoring market at entry level. It was intended as a budget priced new car, but nevertheless, the economies in its production were not reflected by inferior materials but rather an eccentric simplicity. To trace the origins of this car we need to step back to the late-1930s to look at two favourite Ford saloon models of that time, the E93A Prefect and the E04A Anglia (after the war these were modified to become the E493A Prefect and the E494A Anglia). The Prefect E493A provided the engine for the new model, whilst the Anglia E494A donated its body design. The Popular 103E was fitted with an 1172 cubic capacity (cc) side-valve engine from the 10 horse power (hp) E493A, a power unit which was also fitted to the export model Anglia's. The 8hp Anglia E494A was in production from 1948 to 1953 and was, like the 103E, a 2-door saloon. It was powered by a side-valve engine of 933cc which, in turn, had been developed from the power unit in the E04A which first saw the light of day back in 1939. The E04A, although being similarly bodied to the E494A, had a very distinctive rather square bonnet and an almost vertical radiator grill. Incidentally, the E04A model was the first Ford to use the Anglia model name, and it too used the 4 cylinder 933cc side-valve engine as its power-train coupled with a floor mounted gear lever to its 3-speed box Meanwhile, the 1172cc engine that the 103E Popular inherited had previously been used as the power source in the E493A Prefect. This was an attractive 4 door family saloon of 'Six-Light' construction, that is to say three windows along each side of the car. The Prefect's boot lid was flush with the gentle rake of the back of the car, and like most Fords of the period was bottom hinged, a separate panel below allowing access to

Picture Top Left: *A 1946 8HP Anglia two door model, which was primarily designed as a family saloon with a 993cc side valve engine. This was the most popular Ford in the post-war era, and it undoubtedly accounted for many a serviceman's gratuity on demobilisation from the armed forces.*

Picture Middle Left: *A 1949 example of the Anglia marque, this time a model E494A but again an 8HP model. This body style was subsequently utilised in the 103E Popular model of 1953.*

Picture Bottom Left: The *10HP Ford of the period was the E493A Prefect four door saloon. It had a 1172cc engine and was a further development of the Ford Prefect E93A. This post war (1949) version of the E493A shows the modernised integral headlamps and the V8 Pilot inspired radiator grill.*

the spare wheel. At the front of the car access to the engine was gained by twisting the mascot on the alligator-type bonnet. The radiator grill was very similar to that adorning the Ford V8 Pilot, but the head-lights on the Prefect were flush-fitting in the front wings. Incidentally, it was on a certain Rochdale registered Ford Prefect E493A, during one idyllic summer of his far-flung schooldays that Robert Berry learned to hand-crank an engine without breaking his thumb .

The 192,229 examples of the E493A Prefect that were built between 1948 and 1953 (like the E494A) also owed their origins to a pre-war model, in this case the model being the E93A Prefect. This 10hp motorcar had the same 1172cc side-valve engine as the later model, but its origins could be traced back to the power plant of the model C and the 7W of the 1935 to 1937 period. The E93A model Prefects were very similar to their younger sisters; the two models, however, being very easily distinguished by the radiator grill which was of a similar shape, but totally different in its construction. This consisted of fine horizontal convex bars. The front wings were also another point of identification, as these were completely different; although separate of the body in design, the wings of the E493A merged into the bonnet shroud, whereas the wings on the E93A model were of popular pre-war styling and had their headlights mounted on top. Unlike the E493A, the E93A was offered with different bodywork, at least in the pre-war years. The E493A was a 4-door saloon, although isolated examples of commercial 2-door vehicles were to be found. Whereas the E93A was offered as a 2- or 4-door model. Fords actually advertised these models as the Fordor and the Tudor (a simple play on words, Tudor = Two door saloon etc.), in addition to the Tourer or one of the rare Drop-head coupes. Of these models 667 Drop-head coupes are assumed to have been built, as opposed to 1,028 Tourer models, meanwhile 10,170 2-door saloons and more than 100,000 of the 4-door saloon cars are recorded. As mentioned earlier, the E93A and the E04A gave way to the E493A and the E494A in 1948, which themselves were later replaced by the miniature Consul-like 100E models.

Despite Fords extensive modernisation program in the early 1950s, the company must have had a certain fondness for their perpendicular `baby' for she was given more power, dressed up and allowed out to play once more. In 1953 the 1172cc engine was married to the E494A body, then given the name 103E Popular. But as fashions had changed before she left the showroom, the cute little darling was already obsolete cosmetically. As she stood among the Consuls, Zephyrs and newly introduced 100E Anglia and Prefect models, it was like seeing your favourite old auntie on the cat-walk with the top fashion models. Even so, there was an audience there for her and this was the market that Ford was aiming for, as it strove to provide an economical base-entry model. Therefore, for the motorist aspiring to new car ownership, the 103E (Puddle Jumper) excelled herself.

Externally very little different from the E494A Anglia, the 103E used the same 2-door, 4-seat saloon bodywork which was of relatively narrow dimensions. It sat, almost complacently, on narrow pressed-steel wheels; the hubs of which were body-coloured as opposed to having the more usual chrome finish. The doors, like the Anglia, were forward hinged and had a light swage line on the lower area which emphasised the wing line. The side opening bonnet appeared rather high and narrow, and tapered to a fine radiator grill; this grill was comprised of two tall narrow oblongs which swept vertically back from a plain bumper which was painted aluminium colour. Auxiliary grills were fitted on each side of the bonnet sides too, with the bonnet being hinged along the centre-line which enabled it to be opened from either side. Small headlamps were mounted on the rather wide front wings and semaphore indicators were utilised.

Picture Above: *An obvious view of a Ford Popular given away by the painted wheel disks, single windscreen wiper and general de-chroming of the bodywork. This is the 103E model built between 1953 - 1959 and sporting the same 1172cc engine as the Ford Prefect. In time the 103E obtained the nickname 'Puddle-Jumper' because of its tendency to 'skip' when driven with a bit of vigour.*

A high boot protruded a short way from the back of the car with a top-hinged square number plate on the lid. This arrangement allowed the bottom-hinged boot lid (which was retained by adjustable straps), to be lowered for additional luggage capacity and still show the registration plate. Contrasting with the new 103E, the old Anglia E494A had employed chrome bumpers complete with over-riders and chromed wheel hubcaps. It was basically small things like this that differentiated between the two models, but these small savings allowed Ford to sell a more economically priced car. Those who obtained a 103E as their first car remember the model with great affection, although they were hardly the most luxurious of vehicles. Yet the economical price, freedom, and surprising level of reliability did much to enhance Ford's reputation, and this undoubtedly began a breed of customer-brand loyalty which many still retain down to today.

In summary we can readily conclude that with the Popular 103E, the original policy of Henry Ford's 'Motoring for the Masses' was once again realised in post-war Britain. It clearly provided the public with an affordable new car, without which many would have been looking to the second-hand market or, worse still, the back street spivs!

Picture Above: *A Ford V8 Pilot built between 1947 and 1951 the boywork was built by Briggs Motorbodies across the road from the Dagenham plant. Although this vehicle had a V8 engine this was still a side-valve design and had a column shift gearlever.*

CHAPTER 2
THE FORD PILOT V8

It was curious that the Ford Edsel, a car which had been expected to capture the American middle market, became a tremendously embarrassing sales flop. It was aimed at a customer bracket between their Mercury models and the cheaper Fords, but despite all the forward planning, advanced publicity, and the excessive expenditure it failed to achieve its goal. It has often been described as the wrong car at the wrong time. Like the Edsel, which should have been destined for great things, there was a similar story with Ford in Britain with its model V8 Pilot. This was a very British motor-car yet, without being in any way derogatory, it should have been an example of the right car at the wrong time and one which not only survived in the financial jungle of post-war Britain but also collected it's laurels too. However, 1947, with post-war steel shortages, coal shortages and severe petrol rationing was not the ideal year for Ford to introduce a new large car.

It was originally intended for the Pilot to be powered by a new 2.5 litre engine, but there were unfavourable comments about this power unit by some of those testing the car. In those days of severe austerity it was difficult to spend great amounts on refining products, so Ford opted to give the Pilot a proven V8 engine when the main production runs began in August 1947. The Ford Pilot (chassis code E71A for the saloon) made her entrance before the motoring press at the Rembrandt Hotel in South Kensington, London. After being introduced by Ford's managing director, the car was greeted enthusiastically. However, despite this grand introduction, there was not an awful lot that was new in the strictest sense of the word. The body was largely a carry over from the 1937-1939 Ford V8 model 62, although the front end was entirely re-designed. It was given a splendid, pompous-looking, tall radiator grill; this featured narrow vertical chrome bars on a sensitive radii with the discrete V8 Pilot badge above and succeeded the American inspired 'V' shaped radiator grill on the model 62. The front wings were also restyled, but complemented by superb huge chrome headlamps and small side-lights. New too, were the heavy looking fluted chrome bumpers and over-riders. The model 62 had ambled along by the power of a 2227cc side-valve V8 whereas the new Pilot had the 3622cc under its alligator type bonnet; this engine itself had first appeared in 1932 in the V8 model 18 and the model 40 of 1933. The senior management at the Ford Motor Company in England at this time was Sir Roland Smith (managing director) and Sir Patrick Hennessy (chairman). Hennessy had started at Ford's factory in Cork (Ireland) in the 1920s but rose to succeed Sir Percival Perry as chairman, when he retired in 1948. In search for a new large saloon, the management team gave their approval for Briggs motor bodies to construct a few pre-production prototypes of the Pilot as a feasibility study before Ford committed itself to commencing full production. These prototypes performed well, and shortly afterwards full production of the chassis's was started at Dagenham. Briggs continued to play their part, as they fitted bodies to the chassis at their nearby works, and then painted and trimmed the cars before they were returned to Fords for final assembly. What made the Ford Pilot such a good car, was the fact that it had a fine 3.6 engine throbbing away cheerily under the bonnet, even though it only had a 3-speed, column-change gearbox. The engine had plenty of torque, so being a 3-speed was really no great handicap. After all, the cars were used for racing, and various people ran the Monte Carlo Rally in a Pilot, whilst Ken Wharton won the Tulip Rally in one in 1950. This all proved that it was not a primitive antiquarian.

Picture Above: *One of the rally Pilots in the 1949 Monte Carlo Rally driven by J. Reece, it finished 82nd out of 166.*

Picture Below: *This picture shows a model 62 with a 22Hp V8 engine rolling off the assembley line. You will observe the very American-styling of the radiator grill on this car. Notice too the variety of different vehicles in production on the same line.*

The police were keen users of the Pilot too, although the patrol cars came equipped with 12v electric systems instead of the standard 6 volts. Talking of the Police, how many of you recall the little kid trying to sell a head lamp to a street trader in the film 'The Long Arm', starring the late, great Jack Hawkins. In this film a police constable asks the lad where he got a new headlamp from, the lad insists that he got it from a wrecked car and he takes the police officer to look at a Pilot with a crumpled nose!

The engine had a cylinder bore of 77.79mm and a piston stroke of 95.25mm its maximum power was 85 brake horse power at 3500 revolutions per minute, producing a maximum speed of around 87mph. Quite remarkable for an engine that was designed in the early 1930s, and it was certainly no slouch then even if the 0-60mph time of 20+ seconds seems a little laboured today, (my 1970s Opel Rekord still isn't that quick, I measure her in minutes). A Ford Pilot owner, like the one pictured above would expect to average around a miserly 18-20 miles per gallon in fuel which must have caused more than a little concern with petrol being rationed as it was when the model first put in its first appearance back in those uncertain post-war days of 1947.

Internally, the new Pilot had a quality feel about it which was clearly needed as its competition was the Wolseley 6/80, Morris Six series MS, Phase I Standard Vanguard and such like. The Pilot had fully fitted carpets, a heater, a clock, cigarette lighter and provision for an optional radio. Instrumentation was set in a large 'Bakelite' dashboard, and was quite comprehensive. (Export models had a different dashboard, one of aluminium as the Bakelite suffered in hot climates; they had vertical louvres in their bonnet side panels too.) Also on the list of fine things that the Pilot was equipped with were foot rests for the rear seat passengers. The bench seats, almost as comfortable as a living room sofa, were usually cloth trimmed but leather was offered as an option for a little increase in the purchase price. For ease of maintenance, it also had a novel 'Jackall' system, which was a hydraulic jack device that operated on all four wheels. The basic price of the car was £585, but with purchase tax this went up to a staggering £748 5s, and if you wanted a model with leather trim it would cost a princely £764 4s 5d.

In 1948 Ford decided to use the chassis of the Pilot for commercial applications, and it offered the E71C to be bodied by outside contractors. Vans and pick-up trucks were built on this chassis, and it could even be purchased as a very practical and attractive estate car. The E71C left Dagenham as a chassis-cab, but they were developed in various styles and applications by a number of coach-builders; at least five were bodied as limousines and several newspaper companies took van versions. As mentioned above, this commercial version was also developed into several types of attractive estate car, as pictured below; either panelled in steel or as a woody, often referred to at that time as a shooting brake. Perhaps the most famous shooting brake was one which carried the registration mark, VUL 3. It was built with a chassis some 12 inches longer than the standard and was taller too. Built in 1951 for King George VI, it was unique in having a floor gear-change as His Majesty did not care for a column change. The steel-work was painted a dignified green and the radiator and bumpers were black. It spent much of its time on the Royal Estates at Balmoral and Sandringham where it still remains today. For normal customers, when a 'Woody' Pilot was ordered from a Ford dealership it was usually fitted with a body built by the firm of Garners of Acton, West London who (incidentally) built the body on the King's car. Yet, several other companies bodied Pilots too; Ford dealers, W.H.Perry of North Finchley made eight Woody's for example and W.J. Reynolds finished several chassis cabs as pick-up trucks. Production of the saloon version came to an end in May 1951 after 22,155 examples had been built, but a few of the E71C models were continued into 1952 and when production stopped, 3,807 of these chassis units had been made. There had been plans for a new 14hp Pilot but, despite some development and planning, this idea was cancelled with the introduction of the new Consul and Zephyr models that had appeared at the Earls Court show in 1950.

CHAPTER 3
FORD CONSUL, ZEPHYR SIX, ZEPHYR-ZODIAC MkI

Picture Above: *One of the new breed of Ford family saloons was the Mark I Consul, designed for the 1950's with a 1508 cc over-head valve engine. The car seen here is Ford demonstrator UPU 418 and possibly pictured in the Cotswolds.*

As the 1950s began, Ford progressed with its desire to have a range of cars to suit all income brackets from the Popular 103E, through the 100E Anglia and Prefect range to the larger Consul, Zephyr six and Zephyr-Zodiac. Two of these larger cars had been introduced in October 1950 (the 4-cylinder Consul with its 1508cc over-head valve engine and the 2262cc 6-cylinder ohv Zephyr six) in time for exhibition at Earls Court, but it was not until New Year's Day 1951 that production really started. When the range was introduced Ford used the 'Three Box Design' (bonnet, cabin and boot) for the first time in Britain, as prior to this all the earlier models had separate wings. These larger Fords with their full width body shells also had other innovations, as these were of monocoque construction and this offered useful weight savings over the earlier products that had used a separate chassis. Interestingly, they were also the first British cars to employ the Macpherson Strut suspension system; this was a radical change from previous Ford engineering policy and, though it was so revolutionary at the time, its effective simplicity has since made it a widely copied concept. Ford's bolt on panel policy was also clearly evident, with front wings, front panels and the lower half of the rear wing being produced in this fashion; the upper section of the rear wings however, along with the boot sill formed an integral part of the body.

The Consul and Zephyr six were identical from the windscreen to the rear of the car, but the Zephyr six employed a longer front section to house the in-line 6-cylinder engine. Both these models had headlights built high into the leading curved corners of the front wings, with sidelights below, but one of the easiest features that differentiated between the models was the radiator grill. On the Zephyr six this was formed of thin chrome horizontal bars reaching almost to the sidelights, with a radii at the ends;

Picture Top: *Two Mk I Consuls, both of which are later examples of the model as shown by the rear light construction. The picture also illustrates the difference in the registration plate lamp housing; the Consul being body coloured and chrome plated on the Zephyr.*
Photo: Robert Berry.

Picture Bottom: *A Mk I Zephyr Six in a sorry state, awaiting restoration in 1992. Pictured near Dewsbury, in what was probably once finished in the Canterbury Green paint. This particular car was originally owned by the well known theatrical empressarios, the Grade brothers who later achieved greater fame in the television and film world.*
Photo: Colin Moorhouse

the middle third of the radiator grill was also slightly convex and somewhat higher, and reached up into the leading edge of the bonnet; in turn this was shaped to accept the grill on which (incidentally) the red and black Ford badge was centrally mounted. On the other hand, the Consul's grill was a regular oblong of convex vertical slats with a chrome finish; the central bar being much wider to carry the Ford badge. Another distinction was the rear registration lamp housing, for this was body coloured on the Consul but chrome-plated on the Zephyr six. The bumpers were also a little different, as the Consul normally featured plain bumpers whilst the Zephyr six wore over-riders. Both models used the same plain chromed wheel trims with a slightly-raised circular centre section.

During their production years, some of the Zephyr's enjoyed various successful rallying careers, as would the Mk II models that would eventually succeed them in 1956. The Consul's too enjoyed a little of the limelight on the race and rally circuits, but they were obviously over-shadowed by the greater potential of the 6-cylinder cars. The success of these models actually encouraged the Ford Motor Company to set up its own motor-sport competition department in 1952. Ford duly prepared four Zephyr Six saloon cars for the1953 Monte Carlo Rally, which was actually won by a Dutchman named Maurice Gatsonides in a beige Zephyr saloon registered VHK 194. Other successes followed in both works and privately entered Zephyr cars, for example Nancy Mitchell won the ladies cup on the 1953 Lisbon rally and the 1954 RAC rally was won by Edward and Cuthbert Harrison! However it was the production car market which really interested the Ford Motor Company, and in October 1953 the Dagenham plant turned out the first top of the range model in the Consul, Zephyr family. This was the comprehensively equipped, 'Zephyr-Zodiac', which was probably the most glamorous British saloon car of its time. The new large Fords had caused quite a sensation on their introduction in 1950 with their slab sided, trans-Atlantic styling and spacious appearance, but the introduction of this 'flagship' in the Ford range went even further. With its exciting appearance it was, for many, to be the dawning of a new exciting motoring era. The model had a very great impact on the British public and indeed the British motor industry as a whole!

The Ford Zephyr-Zodiac was, as the name suggests, an up-market model of the 6-cylinder car and it contrasted greatly with the rather staid and sombre products that were still the traditional offerings of many British car manufacturers in the 1940's. Internally, Ford continued the glamorous styling effect with two-tone leather upholstery and re-styled door trims with lower sections carpeted to match the floors. There was a vanity mirror on the passenger sun-visor, an electric clock, a cigarette lighter and new head-linings in woollen textile. Even so it was the external appearance which had the greatest effect. The majority of Zephyr-Zodiac's were painted in a two-tone scheme, although the contrasting colours were considered by some to be rather garish. The colours were separated at the waist swage line, where a full length chrome strip was located; a feature which also appeared on the Consul and Zephyr around this time. At the front of the Zephyr-Zodiac, the colours were divided at the point where the bonnet and radiator grill met and, at the rear where the boot lid met the sill. These divisions were at the same height as the side swage line thus giving a perfect harmony of line!

Picture Above: *The Mk I Ford Zephyr-Zodiac range was built between 1953 and 1956 and was almost always in a two tone paint scheme. The paintwork was further highlighted by the wide white wall tyres as shown on this model.*

The badging of the cars was another destinctive feature; whereas the Zephyr had a chrome plated "Zephyr six" cast script on the leading edge of the bonnet, the new luxury model had its Zephyr-Zodiac gold-plated. Whilst the words on the Zephyr were separated by the radiator recess, the Zodiac's script was mounted higher to avoid being obscured by the fog and spot lamps that were mounted as standard on the bumper over-riders. Both 6-cylinder cars had their rear badges on either side of the chromed number-plate lamp housing on the boot, but the 4-cylinder model had the word Consul in chrome fitted on the registration plate lamp housing. The final touches to the Zephyr-Zodiac were a new bonnet mascot, styled rather like a futuristic aeroplane and, most noticeably, beautiful white-walled tyres.

This successful recipe of styling and engineering par-excellence was such, that during the entire production programme of the MkI models, (1950-1956), there were only the most nominal changes. One of these, to the dash board, came in the autumn of 1951 following criticism of a lack of internal stowage space, which led to the dash-board and instrument housing being redesigned, Yet, whilst this new dash-board was on the models at the Earls Court show in 1951, it was not until the following September that it was fitted to production models. The speedometer, fuel gauge and ammeter had previously been in an elongated clover-leaf styled housing, and on the passenger side the same design had been used for a glove-box. However, with the new variant it was deleted in favour of a wide parcel shelf below a new curved dash-board; earlier cars becoming nicknamed "Flat-dash", among old Ford enthusiasts.
From their introduction until the Autumn of 1953, the cars had used the semaphore type of traffic indicator, but after this date all models had flashing lights, white at the front (within the sidelight unit), and red at the rear (though these were later changed to amber lenses). As a matter of fact the entire rear lamp unit (except the indicators) were re-designed the following year, as part of Ford's policy of continual improvement. After the saloon cars had become firmly established, Ford and the specialist bodywork company 'Carbodies' of Coventry turned their attention to a convertible model. After extensive development, 2-door convertible versions of the Consul and the Zephyr six became available, although for some strange reason the Zephyr-Zodiac was not offered in this form. The Zephyr six had a hood which was raised electrically to the De-ville position, but it could be ordered with a manually raised hood if you wanted to save £45. Conversely, the Consul had a manually raised hood, although an optional power hood was available at extra cost.

Picture Top Left:
An official view of a Mark 1 Consul, a model built between 1950 and 1956, and one which changed the styling of British motoring fashion. Along with its brethren in Ford's large car range, it also did much to introduced thousands of Britain's to the joys of 'modern' travel.

Picture Bottom Right:
Even more stylish was the Consul convertible model, which whilst officially produced as such by Ford, was achieved by Carbodies of Coventry who did the conversion with official sanction of the Dagenham works. Pictured here with a closed hood, the car retains its rounded appearance, but with the hood lowered the Consul had a definite American flare and yet was a uniquely British car. For those less warm days, the hood could be placed in the deville position, thus providing a rear windshield but open top.

In 1954 an estate version became available, but to understand how this came about we need to consider the situation facing coach-building firms of that time. Unfortunately, due to the integral body-building policies of most post-war motor manufacturers, companies in this line of business were suffering quite badly, and in turn this resulted in the closure of some very well respected and long established coach-builders. However, the Farnham based firm of E. D. Abbott decided, rather than tolerate the continual squeeze that was affecting their industry, to look around for alternative ways to keep their work-force employed. The managing director Mr. R. G. Sutherland considered that creating an estate car version of an existing saloon car might be an answer! Obviously a great deal of the larger manufacturers were building their own, but there were no estate car versions of the Ford range or the 'E' series Vauxhalls. So choosing the Zephyr six saloon as a base, Abbott's design department produced the 'Farnham' style of estate car. They achieved this by cutting across the roof above the rear window and down the rear roof pillars, a short distance back from the door frames and along the top of the rear wings. Abbott's then welded a new section in its place to comprise the estate 'body', and this was duly fitted with windows that harmonised with the rear passenger doors which had been retained. A new single-piece door (which was hinged at the off-side) was produced for the rear of the estate though, in the interest of economy, it made use of the original rear screen from the saloon. The extended roof structure featured a stainless steel luggage rack with five rubber strips fastened to the roof to protect the paintwork, whilst two luggage strap eyes were located above the slightly raked rear door. Because the conversion was executed in such a way, it both maintained much of the saloon's rigidity and helped to keep production costs down. The completed car was registered OLY 90 and was exhibited on the Abbott stand at the 1954 Motor Show.
Ford gave their full approval of the conversion, and also allowed Abbot's to approach the Ford dealerships.

When demand for this conversion proved encouraging the 'Farnham' was offered as a conversion on existing Consul, Zephyr six and Zephyr-Zodiac saloon cars. Interestingly, the service was not only offered on new cars, but as a complete conversion on existing vehicles. However only a small proportion of the extensive production range were ever produced as anything but saloons and accordingly such conversions are now extremely rare.

By far the rarest version of the Consul and Zephyr range was designed and built at Ford's Geelong plant in Victoria, Australia. To suit the needs of the vast open spaces of this continent, the enterprising staff at the Ford plant came up with the idea of a pick-up truck or 'utility` as they call them down there, unfortunately Dagenham did not have the same enthusiasm and the production figures reached double figures, but only just: a mere 11 were made before the project was ended! The last of the Ford Consul, Zephyr six, Zephyr-Zodiac, came off the assembly line at Dagenham, on 22nd February 1956, the MkII models having gone into production the previous day. Yet, during the production years of the MkI models, 231,481examples of the 4-cylinder Consul (chassis code EOTA) were built and 175,311 of the 6-cylinder cars (chassis code EOTTA), of which 22,634 were the luxury Zephyr-Zodiac.

Picture Top Right:
Seen in 1985, a rare survivor of the Mk1. estate range, a Consul model is pictured in a vehicle rally at Buckingham.

Picture Bottom Left:
Another of the Carbodies models, this time a Zephyr convertible, interestingly the Zodiac was never given such a conversion.

CHAPTER 4
THE 100E SERIES

Picture Above: *Similar in styling to the Mk1. Consul, Zephyr, Zodiac range, the smaller 100E series (illustrated by the Anglia model pictured above) was an instant 'hit' with drivers who wanted modern styling and a lower price.*

Whilst we have so far discussed the larger car market in the early 1950s, the growing need for a moderately priced family saloon car was also becoming increasingly evident in this period. In response to this market trend, Ford introduced its economical 100E range of small cars which embodied modern methods of construction. Because they were built along the same principles as the Consul and Zephyr models, the body styling was in many ways rather similar although the dimensions were more compact. A standard body shape was employed for two fairly similarly 100E series cars in 1953, these being the four-door Prefect and the two-door Anglia. The overall dimensions were identical, with a total length of 12' $7\frac{1}{4}$" on a 7' 3" wheel-base: the height was 4' $9\frac{3}{4}$" and it was 5' $0\frac{1}{2}$" wide. Yet, even these compact dimensions were neatly disguised by the attractive body styling. Both models were of conventional front engine - rear wheel drive layout, with a side valve engine. This was a four cylinder 1172cc unit, with a bore and stroke of 63.5mm x 92.5mm.

The 100E models were all driven through a gear-box with three forward and one reverse gear, which had a floor mounted gear lever, and only body styling differentiated the various types. The Anglia was furnished with a radiator grill of three slightly curved steel bars, finished in an aluminium paint. On the other hand, the Prefect's grill was constructed from thin vertical slats in the same general oblong shape, but it had a gentle convex from top to bottom. Both models had a forward hinged bonnet, with the respective name on the leading edge. Headlamps were set in the top of the wings, with a rectangular side-light/indicator below. On the Prefect the lamp bezels were finished in chrome plate, but on the Anglias they were painted in the body colour. Similarly the Prefect had chromed bumpers, whereas aluminium paint was employed on the earlier Anglias. The wheel arches had a slight lip around them and the rear ones were noticeably lower, which tended to lengthen the style of the body in an appealing way. Finally, the 100Es had plain, chromed wheel discs with a central pimple, very similar to those on their larger brethren the Consul/Zephyrs.

Improvements to the range came first in 1955 with the introduction of a deluxe model, followed by styling changes to the standard models in 1957. New, larger rear-light clusters were introduced, and all models were then provided with the larger rear window to improve visibility. The Prefect lost its winged bonnet mascot which was replaced with the smooth, swept 'V' ornament that thereafter adorned all the 100E models. Deluxe models had a thin chrome-plated strip along the waist swage line, and chrome inserts to the headlamp bezels. In July 1954, Ford had expanded the 100E range with the introduction of the 300Es. This was a range of light commercial vehicles, with a 5cwt van which had the Anglia's front end styling. The following year, a 7cwt van was introduced and this employed the Prefect radiator grill. Also in 1955, on 23rd September, the Squire Deluxe was announced as the estate version of the van which featured Prefect appointments.

Picture Above:
The 100E model Prefect, a 4-door saloon with a distinctive grill of vertical slats and chrome headlight surrounds. Its styling and origins can easily be demonstrated by turning back to look at the previous pictures of Mk1 Ford models.

Picture Below Left:
A model 100E Popular De-luxe, 2-door saloon, showing the circular tail-lights unique to this model.
Photograph: Robert Berry

For the first two years of its life, the model featured a little wood trim along its flanks - perhaps aping the American station-wagon vogue. For the more economy-minded, Ford introduced the Escort Estate a few days after the Squire - this being a less well-equipped estate version of the Anglia. The Escort had the same three bar radiator grill as the Anglia but, from 1957 onwards, both models featured a re-designed light-weight grill constructed of horizontal and vertical cross mesh. On the standard models this was finished in aluminium paint, though the deluxe model was given the luxury of a chrome finish.

The production of the Squire Deluxe came to an end in 1959, whilst at the same time the 100E Prefect was phased out. A replacement for the 100E Prefect was found in the 107E Prefect, which although bodily identical had a number of major differences from the earlier model. The most significant was the use of the new overhead valve engine and drive train which had been developed for the new 105E (rake-back window) Anglia. Externally, the only difference between the two Prefect models was the chrome-plated waist line which now came back from the front wing to the wheel arch in a 'V'. This strip was used as a dividing mark for the pleasing two-tone paint schemes carried by many of the cars. However the 107E was only intended to have a short life, for in reality it was only a stop-gap measure to fill the vacancy of a medium range car until the Consul Classic was ready to go into production.

With imminent launch of the new Anglia and Classic models, with their controversial backward sloping rear windows, the end of the 100E range was signalled. The Anglia 100E was replaced by the Anglia 105E in September 1959, and in a sense this paved the way for the Popular 100E. Following their success with the economical 103E in 1953, Ford once more employed the concept of using a superseded model to produce a budget-priced family car. The result was basically a de-trimmed Anglia with the cross-meshed grill of 1957 vintage, however the tail lights were different items - being separate circular units with a round reflector above in a vertical plane. These 100E Populars were available in standard or deluxe versions, but the standards were very basic, lacking such items as door pulls, interior lights, passenger sun visor, opening quarter lights and ash trays - all of which featured on the deluxe version. The Escort estate was still available until 1961, and the 100E Popular remained in production until the summer of 1962, when the era of the small side-valve Fords came to its final end but the body shape went on to house the OHV engine in the 107E mosdel.

The van bodied version of the 100Es, more correctly known as 300E series. The 7cwt Thames van pictured top. Meanwhile the Squire De-luxe estate is seen bottom left with the earlier wooden trim (a feature deleted from the range in 1957) and the Escort Estate with its earlier slatted grill. Again, a 1957 styling change resulted in this feature being changed.

CHAPTER 5
CONSUL, ZEPHYR, ZODIAC MkII

A Comparable view to that used on the front cover, and showing a Mk II Zodiac saloon shortly after introduction in 1956. The absence of chrome trim from the rear roof pillar and the car's circular side-lights betraying its age. The picture below shows the post-October 1956 model, another 'highline' car.

As the decade progressed Ford's large car range was completely re-designed in 1956, and shown to the public in February of that year. The MkII Consul, Zephyr and Zodiac models were bigger and better than the earlier models which thereafter became retrospectively known as the MkIs. Not only were they bigger in almost every dimension, but they were also better in many ways. Ford re-designed many of the component parts, improved the braking systems and gave a better weight distribution. But it was at the heart of the car, the engine, where some of the most significant mechanical changes were made. Although the 4-cylinder Consul shared the same bore dimension as the 6-cylinder Zephyr-Zodiac cars, these were increased to 82.55mm x 79.5mm. In turn this increased the swept volumes to 1703cc on the Consul, and 2553cc on the other two models. A two-door convertible quickly followed the saloon car into production, and towards the end of 1956 an estate version was once again offered, with the conversion by E.D. Abbott of Farnham. In many ways this was identical to the conversion on the MkI estate models, but there were a number of subtle differences not least of which was the window-line.

Pictured Above:
The Consul variant of the Mk II, but somewhat a hybrid model as it employs rear wings of a highline and a lowline roof panel and lamp bezels - dating this as an early 1959 production.

Pictured below:
A 'low-line' Zephyr model pictured at the rear of Beaumont Brothers Garage, Meltham, West Yorkshire in the company of a Morris 1000 convertible. Photograph Alan Earnshaw

The new saloon cars, meanwhile, followed the same basic design principles of the MkIs, but they had a completely fresh style with an up to the minute trans-Atlantic flair. They also had an air of beauty and grace, and as modern-day advertisers might say, 'more than a little panache'. The bodywork on all three models was almost identical from the windscreen backwards. On the Consul there was a subtle curve to the top corner of the rear wings, on which they mounted the rear light units - these were a triangular amber indicator at the top, a rectangular rear light in the centre, and a small round reflector below. This arrangement differed from the Zephyr-Zodiac models which had an identical rear wing, but had a more squared-off style, though each model again had different light clusters. The Zodiac, which in MkII form was a different model in its own right, also had a ribbed metal trim on the rear panel below the boot lid which embraced the rear wings just below the rear lights. The rear registration plate was hinged and sprung at the bottom, allowing access to the fuel filler cap. All three models had bolt-on front wings, though as before there was a difference in the length between the 4-cylinder and 6-cylinder models. From 1956 onwards, the whole of the rear wings became part of the body structure on the MkII. The front panel below the radiator grill was also made as a bolt-on item, and the bonnet and boot were then counter-balanced. However, it was the panoramic views afforded by the new body-styling which figured as the major feature on the MkIIs. From the driver's seat, the forward view was excellent with the bonnet sweeping gracefully down between the wings. The style of the wings had the added advantage of making manoeuvring the car accurate and almost effortless, even in the most confined spaces. The rear-screen had the same panoramic features, with the pillars set back so as not to impede the driver's field of vision. The rear view took in both of the rear wings and, because of this, even the reversing of these large cars was an easy exercise.

The major identifying feature of each model was the radiator grill, with all three models being different. On the Consul it was a wide oblong with a mesh effect, the top of which had a gentle curve from end to end with a separate trim section that followed the curve. The Consuls also had a chrome-plated trim item, in the form of a narrow casting which swept round the bottom corners of the wing just above the bumper. The radiator grills on the Zephyr and Zodiac were of a similar fashion to those carried on the MkI models, but were wider and more streamlined. Conversely, the Zodiac grill was formed of thin horizontal bars, with a small casting of fine vertical bars above the central section. Zephyr grills were a one-piece structure of the same overall shape but constructed in an elongated waffle shape, although this was re-designed in October 1957.

The MkII Zodiacs were often finished in a two-tone colour scheme, separated at the chrome waist band, giving an imposing impression when shod with white-wall tyres, as we show on the front cover of this book. Originally there was a choice of five two-tone paint schemes or black for the Zodiac, and eight monotone colour schemes for the Consul and Zephyr models. These were Black, Sarum Blue, Carlisle Blue, Wells Fawn, Ivory, Warwick Green, Corfe Grey and Hereford Green which is seen in colour on the back cover of this book. Interiors were also usually two-tone, for example a black car would have red/cream interiors. Early cars had buttoned vinyl seats. In the autumn of 1957 Ford continued its expansion into the quality car market when the Consul Deluxe was introduced, with interior appointments almost to the specification of the Zodiac models. Externally the deluxe models were distinguished by a second colour applied to the roof panel, and by chrome-plated rear lamp clusters.

Pictured above:
Lovingly restored, this Mk II Consul is pictured outside the Cumbrian home of author Robert Berry who was instrumental in rescuing it from a Yorkshire barn. Restoration was completed by Colin Moorhouse.
Photograph Robert Berry

Pictured below:
The Mk II Consul Estate, converted by Abbots of Farnham who used as much of the original body shell as possible with the sanction of Ford.

Estates Galore

Viewed in the picture left, a restored Mk II 'lowline' Consul Estate looks as glamorous in the 1980s as it did when built 25 years earlier. Above is a 'High-line' Zephyr estate which dates from around 1958, whilst below a 'lowline' Zodiac estate is seen in Bradford in the late 1980s. The poster advertising the new Ford Sierra, seen in the background makes an interesting contrast. Little wonder that, as this book goes to press in 1997, Ford can rightly be celebrating being the nation's favourite car manufacturer for 21 years.

Photographs: Robert Berry Collection

Following the introduction of this model, the paint colours were revised so that early in 1958 you could have, Kenilworth blue with a black roof, or Pembroke Coral with a Dover White roof, alternatively there was a Dover White body with a Ludlow Green roof, whilst Pembroke Coral also appeared as a roof colour on Durham Beige coloured cars. Though this continual revision of colour schemes was an on-going process, it was apparent someone at Ford had a fondness for place names.

In February 1959, three years after introduction, a modernisation programme was carried out to take the range into the 1960s. All models were given a lower, flatter roof panel and chrome-plated headlamp bezels. The thin chrome-plate waist strip now ran all the way to the end of the rear wings, whereas before it had stopped short on the Consul and curved upward on the Zephyr. The Consul's rear wings were then squared up, and the deluxe version had chrome-plated strips added to the four horizontal ribs of the back panel. All three models received new rear-light clusters, and wider, brighter window trims. The interior was re-designed with a padded fascia and a lower instrument cowl replaced the

semi-circular type in earlier models. Newer, brighter colours were introduced, and the estate cars were then available in either a two-tone or monotone finish. The colour division (depending on a new chrome body-strip that curved down from near the front windscreen) then ran parallel with the waist strip to the end of the vehicle. Zodiac estates were also available in a two-tone scheme in a sandwich effect, in addition to the standard two-tone and monotone schemes. All modernised MkII cars then became known as 'Lowline models' whereas the pre-1959 cars were referred to as 'Highline models'.

In 1960 all three models became available with front disc brakes as an optional extra, as opposed to the standard drum brakes operating on all four wheels; but the following year disc brakes themselves became standard. Also in 1961, the Consul was re-named the Consul 375 to differentiate it from the new Consul Classic. The Consul 375 logo then appeared in chrome lettering on the right-hand corner of the boot lid and in the radiator badge. In the spring of 1962 the Consul, Zephyr, Zodiac models were deleted from the Ford range, with the MkII models having been produced for six years and two months.

Besides the saloons, convertibles and estate cars, the Dagenham plant also manufactured a hand-full of pick-up trucks. However these differed greatly from the pick-ups and estates manufactured at the Australian factory which were aimed at the peculiar Austral-Asian 'utility' market. A good many of the British pick-ups found employment with small businessmen, but because of their power they became exceptionally popular with small farmers who used them for milk deliveries and general farm work. The MkII saloons were very popular with a large number of police forces, both home and abroad, whilst they were also held in high esteem by a number of rally drivers. Though in the main it was the private motorist who ensured the success of the range introduced as the 'Three Graces' when they were introduced in 1956 - a description in which they excelled.

Contrasting Convertibles:
In this sequence of pictures we show the Mk II range of convertibles, all of which were produced from standard Ford saloons by Messrs. Carbodies of Coventry. Featuring the Consul model top, the Zephyr model centre, and the Zodiac bottom. The firm of Carbodies were based in Coventry and produced conversions for a variety of manufacturers, but it was with their work for Ford that they achieved their greatest glory. Their products featured regularly at the Earls Court Motor show, and due to their stylish and sporty look they achieved a significant number of sales amongst the discerning motorists. For many others they were as much a fashion statement as the cabriolet cars of today.

Even down to today these cars have a remarkable pulling power, and some Ford dealers still have them in their showrooms to attract the public to more modern cars. One such dealer is the firm of Beaumont Brothers, whose garage was built on the site of the stables once owned by the Earnshaw family haulage and carting business at Meltham in West Yorkshire.

CHAPTER 6
THE ANGLIA 105E/123E

Pictured Above:
Typical of Ford's publicity pictures of the 1960s, this view of the Anglia 123E (1200cc) saloon, demonstrates the 'family attraction' of the small cars which Ford were producing for the new era.

By the latter part of the 1950s Ford were in desperate need of a new car with modern styling to capture the small car market of the 1960s. At that time most of the offerings from British car manufacturers were along the same lines, with round bulbous features. The Morris 1000, Austin A35, Hillman Minx, and Ford's own 100E range all falling into the same concept. Much the same was evidenced in the European market, with perhaps the bubble-shaped Volkswagen Beetle being the prime example. Something was needed to break the mould, and Ford intended to do it by employing American styling on a small European car. During the latter part of 1958 a new, much disguised, small Ford first appeared on the roads and motoring correspondents got a hint that something revolutionary was about to appear. However it was not until the Anglia 105E appeared in Ford dealer's showrooms around the country in 1959, that the radical new style of the car became widely known. It was quite revolutionary, with a completely new body and crisp angular lines that provided a compact 2-door - 4-seat saloon.

It was also equipped with an entirely new engine design, although this and the drive train were features often overlooked in the press. Perhaps the lack of attention to the power unit was because of the radical styling, but the engine was the heart and, in many ways, the success of a car which was to have nearly a nine year production span.

The Anglia had a 4-cylinder overhead valve engine with a greater bore size than stroke, which enabled the car to have higher cruising speeds and belied the fact that its engine size was only 997cc. Even so, it was the body styling which featured so prominently and gave it what several commentators called 'a shark-like appearance'; primarily because the backward rake of the rear-screen looked like the dorsal fin of one of the denizens of the deep, though even the radiator grill continued the impression, with an almost mouth-like appearance. With my parents owning an Anglia from 1962 onwards, I never saw such features in the car - but my wife (whilst still a schoolgirl) was knocked down by an Anglia 105E in 1964, and always maintained they had a vicious look. Be that as it may, the styling caught the public imagination and a place on an extended waiting list became the situation many would-be owners found themselves in.

Headlights were mounted high into the top leading edge of the front wings, set under a slight cowl. The bonnet swept gradually down between the wings on to the top of the low, wide radiator grill. On the deluxe model this grill was a chrome mesh structure that extended to the triangular-shaped indicators at the very front corners of the wings, just above the bumpers. On the standard models, the grill was of thin, horizontal, body coloured slats which did not extend as far as the deluxe grill. The rear wings were rather square, and though not amounting to the finned versions of early-60s American Fords, they gave pretensions towards these in their appearance. This was accentuated by the downward sweep of the boot which dropped between the two rear wings to meet a rather high sill which housed the number plate, and (offset to the right) the fuel filler. The rear of the wing 'fins' also housed the tail-lights, but again it was the backward rake of the rear-screen which suggested the non-European styling.

Picture Top Right:
Certainly less glamorous than the previous model, the 105E in this picture is seen outside the David Brown Tractor factory near Huddersfield, with a youthful Alan Earnshaw at the wheel.
Photograph: Ellen Earnshaw

Picture Middle Right:
A late model 105E Anglia dating from 1967, this time a De-luxe version seen in Huddersfield near an early model Escort, the car which superceded the Anglia.
Photograph: Robert Berry.

Picture Bottom Right:
Just right for those family days out, this Anglia 123E is seen at Carlisle Races by the two furlong post. The popularity of the model influenced a generation of drivers, many of whom went on to owning Escorts or Cortinas.
Photograph: Alan Earnshaw

Basic models were painted in monotone colours, but the deluxe version was usually painted with a differently coloured roof, in addition to a thin chrome-plated strip along the waist. In August 1962 a 'Super' version was introduced with an 1198cc engine and provided with almost identical appointments to the deluxe, though externally there was one major difference between the two models. The new 'Super' was also finished in the two-tone paint scheme as applied to the deluxe, but it carried a full length tapering flash along the waist. This was usually finished in the same colour as the roof panel (though not always), and boarded by thin chrome strips. The model was also provided with badges displaying '1200 Super' on the boot lid. With the introduction of this more powerful version, the differentiating terms of 105E and 123E were applied, the larger engined car being the 123E.

The Anglia range was very popular from the outset, and it was not surprising that variations of the saloon became offered to an eager market. First came the estate cars which were produced from 1961 until the range finally came to an end in 1967/8. This was then followed by a small commercial van which found a good deal of popularity with small businesses and fleet operators alike - including ice-cream-men, who can fail to remember the large fleet of Wall's vans for instance. Less popular was the pick-up version and only a few of these were manufactured, whilst the Anglia convertible was almost a non-existent species. One final and very unique version of the Ford Anglia was only encountered abroad and even then, only in very small numbers. This model was the Anglia Torino and was apparently developed at the suggestion of a certain F. M. Paradise who was the head of Ford in Italy. Sales of the 105E had at first been very good, but within a couple of years the Anglia had totally lost appeal with the tremendously fashion conscious Italians. So Ford agreed that a new shape could be produced in Italy on the provision that, as a cost-saving measure as much of the Anglia as possible be retained and incorporated into the new design. The responsibility for the new design was given to the renowned stylist, Michelotti, who submitted plans for a very fashionable vehicle. While retaining the windscreen and doors of the 105E, it was totally different forward of the screen and aft of the doors. The front of the car had a regular box-like profile with a level bonnet and wing-line, the radiator grill being of seven fine chrome bars with circular head-lights set at each end of the radiator, the whole being surrounded by a thin chrome band with small rectangular sidelight and indicator light units below the radiator grill but above the plain bumper.

During the production span of the Anglia, work was transferred from Dagenham to Halewood on Merseyside. Exactly where this picture was taken is not recorded, but it shows a fascinating view of body-shells being run through production.

The controversial rear screen was changed for one with an outward rake, giving a larger side window too, while the boot itself was very similar in style to the Triumph 1300, but utilising circular rear lights. Production of the Anglia Torino was undertaken at the OSI works in Turin between February 1965 and 1967. The cars were sold in Italy, Holland, and Belgium but never officially imported into Britain. As far as the home market went, and despite the Anglia's popularity (as demonstrated by the continuing waiting lists), Ford decided to supersede it with a new model in 1967, when they almost at once reverted to a semi-bulbous shaped small saloon. Everyone thought the model was doomed to failure, and there were few waiting lists for the new car, but it surprised everyone and it is still with us (after various transformations) 30 years on. It was of course the Ford Escort. Ford moved boldly with their new car, and aimed at capturing the 'F' registration market when, for the first time since 1963, the registration suffix letter moved to commence on 1st August instead of New Year's Day. The Anglia models were quickly phased out, but rather surprisingly the later models were turned-out in either a metallic gold or metallic blue finish. Among the last of the Anglias to be built were a few estates, the production of these being continued until the Escort estate was ready. Therefore a number of Anglia estates were still being registered as new cars with the 'G' plate of August 1968. Likewise the production of the van continued into 1968 when they too were finally replaced with an Escort model which, like the Anglia, is worthy of a book on its own.

Pictured Above:
An Anglia Estate in an official view of the early 123E model.
Pictured Below:
The commercial 'Anglia', the Ford Thames 307E 7cwt van.

CHAPTER 7
FORD PREFECT 107E.

Pictured Above:
Basically brilliant, the 107E was not the most luxurious car for the 'swinging sixties', but great for the economy minded motorist, or those wanting to upgrade from a motor-cycle or public transport.

With the introduction of the Ford Anglia 105E in 1959 to succeed the 100E models, the Ford Motor Company, again appeared to have something on their hands destined to be another runaway sales success. With its modern, crisp, clean-cut lines, that new 997cc over-head valve engine, and now a 4-speed gearbox too, how could it possibly fail? But Ford's marketing department were not quite satisfied, because the body shape of the Anglia did not lend itself to a 4-door configuration. Sadly sales of competitively priced 4-door saloons such as the Morris 1000, Austin A35 saloon and the like were taking sales away from Ford to BMC. Triumph actually had a similar problem with its 2-door Herald, but Ford felt something had to be done. True, the planning department had something to offer very soon, but at this moment in time Ford only had the 100E Popular, and the new 105E Anglia (both being 2-door saloon cars) in the lower price bracket. For anyone who wanted a 4-door Ford, it was quite a climb up the financial ladder to achieve the level of the Ford Consul. More importantly, not everyone wanted a car of the Consul's dimensions or that size of engine. The solution was, in the interim, to continue with the 100E Prefect body-shell but to up-rate this in performance terms with the new engine, running gear and gear-box of the 105E Anglia. In short, Ford had come up with a winner again, providing a new budget-price 4-door saloon car without further development costs and the associated teething problems which all new cars face. Given the model code 107E, the new Prefect with its 4-speed gearbox became a firm family favourite. It made driving around town so much easier, and thus encouraged many ladies to take out their first license, but even so it had a livelier, more complacent feel. Fuel consumption was also improved, with an average of around 35 m.p.g. as opposed to something like 30 in the 100E. The body was still the same, with enough accommodation for four adults though it would take five at a pinch.

Externally the 107E differed very little from the 100E, but it featured the larger rear screen and new rear lamp units of the later 100E Prefect. It also had the same frontal treatment, but this time the 107E was finished in two-tone colours separated at the waist swage line by a full length chrome strip. However, before reaching the front light unit, this piece of body trim came back on a 'dogs-leg' to the front wheel-arch, giving a forward thrust to the lower colour. The 107E Ford Prefect was manufactured from October 1959, and deleted from the Ford range in March 1961 after 38,154 examples had been produced. Its natural successor was the Classic, and later still the fabulous 4-door Cortina which epitomised Ford's most successful entry into the medium-priced family and commercial-traveller's car market.

Picture Above:
This official view of the Ford Classic 4-door saloon from 1962, shows to great effect the controversial styling applied to this model. Initially plagued by several power unit problems (which undoubtedly shortened its life-span), it was still a good looking family saloon and was just right for the 'sixties'.

CHAPTER 8
CONSUL CLASSIC AND CONSUL CLASSIC CAPRI

Following quickly on from the rake-backed window of the Anglia, Ford decided to continue this styling concept in its new intermediate size car the Consul Classic. Placed neatly between the 997cc Anglia and 1703cc Consul 204E, the Consul Classic 315 came onto the motoring scene with a 1340cc OHV engine in 1961. Aimed squarely as a smart family car it could seat four to five people in comfort, and there was a choice of 2-or 4-door models - the latter being most urgently needed as we have already seen. The 4-cylinder engine was basically an enlarged version of that used in the Anglia, whilst the 4-speed gear box offered an optional floor mounted gear lever to the standard column change.

For many the backward sweep of the rear screen was the car's most pronounced feature, but the Classic also had a number of other novel styling features. Four headlights were incorporated into the design, with the wing and bonnet line slightly hooded over them. The bonnet was forward hinged, with the central portion sweeping down to a very low radiator grill between the headlights. The grill had five, four-pointed stars set within, and was mounted directly above a plain bumper, beneath which was a further portion of the radiator grill in the form of a small crescent shaped area. The profile of the Classic was designed in the American vogue to reflect speed-styling, a feature achieved by subtle use of swage lines and body panel pressing. The main swage line started at the front wing and swept just below the windows towards the rear of the car, progressively deepening until it formed a modest fin over the rear tail-lights. A very light swage line was featured a little further down the car's side, and a noticeable lip was formed around the leading edge of the

Pictured Above:
Once again outside the David Brown Tractor factory, the young lad climbing into his father's new2-door Classic (purchased at Red Sign Garage, Mosley) was to grow up to jointly write this book, and one on the tractors produced opposite his house.
Photograph: Ellen Earnshaw

Picture Below:
Another Huddersfield view of a Classic, taken this time outside the Friendly & Trades Club, showing a four-door saloon, in Imperial Maroon and Dove Grey.
Photograph: Robert Berry

rear-wheel arch and ran to the rear of the car. All this further serving to accentuate the 'accelerated' look. Quarter-lights were fitted to the front side windows, whilst the rear side windows were styled to harmonise with the rear rake of the car whereby the top of the rear-side window curved down slightly, while the bottom curved rapidly up to meet it. The boot was quite large, and ample enough to meet the load of an average family. Its lid was spring-loaded, and hinged very close to the bottom of the rear screen. The boot also housed the spare wheel, stowed in an up-right position within the left wing.

Internally the car was appointed with the luxuries to be (then) expected from the Consul range. The glove-box in the fascia was fitted with a lock, whilst there was plenty of additional space on a full width parcel shelf located below. Padding was placed above and below the fascia, in an attempt to minimise injury in the event of an accident. It worked, and worked well as I can testify having bounced my head off of it when colliding with a fence at around 60mph at Leicester Forest East Motorway Services in 1969. Directly in front of the driver, and beyond the twin-spoke steering wheel, a deeply recessed quadrant speedometer was located: this instrument panel was also furnished with a fuel gauge, and a water thermometer. De-luxe versions of the car featured additional instrumentation as well as a full-circle chrome-plated horn ring on the steering wheel. The windscreen wipers were now fully electrically operated, as opposed to the traditional vacuum drive long associated with Fords (though many of the early Classic models were plagued by repetitive failures to the wiper motors). Another unusual feature of the wiper system was that the wipers were actually twin-bladed.

The Consul Classic had a dry weight of 18cwt, a turning circle of 34ft, and overall dimensions of 14' 2¾" long, 5' 5½" wide 4' 6¼" high, with a 8' 3" wheel base and a 4' 1½" track. Braking was hydraulic, using the Girling system with 9.5" discs at the front, and 9" diameter drums at the rear. Suspension was again with the Macpherson strut system at the front end, whilst semi-elliptical leaf springs were found at the rear. Steering was by Burman Re-circulating Ball and a three-piece track rod. All of this should have made a car more than satisfactory for the needs of the average family, but something was sadly wrong, and the Classic seemed plagued by persistent engine defects which received much adverse comment in the motoring press of the day. By 1962 the situation had become so bad, Ford announced that it was introducing a new, re-styled 1500cc engine. Those of us who remember this period, whilst recalling the frustrations will also remember Ford's customer care policy which was generally so rare in those days. When a car had to be returned for rectification it was not uncommon for a loan vehicle to be made available for a prolonged period, whilst colleagues of mine recall that they were actually provided with rail tickets to get them to work whilst their car was being recalled for rectification.

By 1962, with most of the problems behind them, the Classic began to capture its deserved share of the market. The standard model was available in ten single tone colours, Ascot Grey, Windsor Grey, Savoy Black, Aqua Blue, Ambassador Blue, Lime Green, Goodwood Green, Panama Yellow, Ermine White, and Imperial Maroon. As for the deluxe version, it was available in six duo-tone colours, with the roof being different to the body colour. However, in the case of our second Classic, we had one of the rarer models which had both the roof and side flash painted in contrasting shades. The body colours were, like the Anglia de-luxe, Windsor Grey, Imperial Maroon or Ambassador Blue (all with Ascot Grey as the contrast colour), or Aqua Blue, Lime Green, or Ascot Grey with Ermine White as the contrast. As part of the drive to regain customer confidence, Ford improved the range with some subtle changes and in 1962 the 1340cc version became the 109E and the 1500c engined cars became the 116E. No estate cars or convertible versions were ever officially produced, although there were unofficial conversions. Sadly, the Classic had not been the success Ford had hoped for and this new flagship of family motoring was destined to flounder by the end of 1963 in preference to the Cortina which was rapidly developed to use both the 1500cc engine and a number of other components from the Classic. In the end only 84,694 1340cc versions were built, with even fewer (26,531) 1500cc models taking to the road! We had both versions in our family, along with a Lotus model (UVH 21) - they are sadly missed.

Pictured Above:
The official pre-launch photograph of the Ford Consul Classic, supplied by the company's library at Averley. For her help in assisting in the location of the photographs used in this book, the authors would like to acknowledge their grateful thanks to Fran Chamberlain. It should be said, however, that the pictures in this library are not generally available to readers.

Consul Classic Capri

Pictured Above:
This Essex registered Consul Classic Capri is another Ford demonstrator, and the picture is obviously posed to reflect the luxurious life-style that this car would introduce its owners to. The drinks would, if unused, be quite vintage too by now - a pity the champagne had been opened.

The Ford Classic saloon had only been in production a few months when the company's range was further expanded by the introduction of a smart little sports coupe version. It was something of a departure from traditional Dagenham products, for never before had Ford of Great Britain produced what can only be described as a 2+2 seater coupe. The mechanical specification was almost identical to the Classic saloon, and the few variations that were evident were of an extremely minor nature. It was the body styling that differed, but even then the overall weight was around the same 18cwt mark. The improved air-smoothed shape did give the coupe an edge of an extra one or two miles per hour over its sister, along with marginally better fuel consumption. Below the waist line, the cars were again almost identical with the 2-door version of the saloon, but it was above where the major differences were immediately noticed. Both cars had the double curvature to the windscreen, but the Capri had a lower roof-line that swept gently down to a rear screen which was both wide and curved. This screen was set at an angle of some 40 degrees to the horizontal, thus making it a complete contrast to the rear-sweep of the back windows on the Classic. It was possible to have a long open area to the side of the cabin, as not only did the door windows wind fully down, so too did the small triangular windows to the rear of the door.

The windows actually wound down in a clockwise or anti-clockwise direction depending on your point of view, so that the vertical edge/frame of the window lay horizontally along the lower window sill, leaving just the small quarter-light on the door showing. The roof line of the car was a good two inches lower than the Classic saloon, a feat achieved by interior alterations and a lowering of the front seats. There was a bench-type seat in the back, which was barely sufficient for two adults although it was quite satisfactory for small children, but not much use for courting in! As the Capri was intended as a 2-seater, the back bench was just that, although the manufacturers did offer the 'luxury' of a cushion to cover the platform as an optional extra.

The Capri was only available in the deluxe version, and as a result the coupe had some luxurious appointments. Over-riders were usually fitted as standard to the bumpers, whilst bright wheel trims and white-wall tyres were available as an extra. The interiors were fitted with white PVC head-lining covering a fibre-matting sound-proof material. The floor was fitted with a bri-Nylon carpet and the front seats were larger and contained more padding than those in the Classic. Some were trimmed with leather, but the majority were in two-tone PVC to reflect the car's external colour scheme. The seats were forward hinged, thus allowing access to the confined back 'seat' area. Like the Classic it was also re-equipped with 1500cc engine in 1962, after which the 1340cc version went out of production, the total number built being 11,143. The 1500 stayed in production until 1964, during which time 7,573 models were built. A 1500 GT model was soon made available, but only 2,002 cars of this high-performance version were produced: however, those which took to the road were capable of upwards of 95mph in the days before the national speed limit was set at 60/70mph.

Picture Above: *Around 10 years old at the time the picture was taken 315 HAJ is seen at Bradley in West Yorkshire. By this time the new breed of Capri had already come on to the market. These modern Capri's, although bearing the same name, had very little in common with the Consul Classic Capri. However these cars themselves have gone on to become highly sought after classics and have a wide following amongst Ford enthusiasts. By way of comparison the Consul Classic Capri has survived, but in much fewer numbers.*
Photograph: Robert Berry

Picture Left: *Another original publicity view of the Consul Classic Capri taken around the same time and location as the previous shot. Precisely where this picture was shot is not recorded in the Ford Library, but it is believed to be alongside the River Thames in Berkshire. As with all the* **Nostalgia Road** *books, the publishers like to use entirely period pictures, so we would be delighted to hear from readers who can help identify the locations of the unknown views used in this book, as well as from anyone who would like to contribute illustrations of any sort for future publicaton.*

CHAPTER 9
FORD ZEPHYR 4, ZEPHYR 6, and ZODIAC MkIII

Pictured Above:
Again representing the luxury life-style that could come with Ford ownership, this picture shows a left-hand drive Zodiac Mk III outside the Cassino at Monte Carlo.
Pictured Below:
Perhaps a more representative view of a Mk III in Britain

Although the MkI and MkII Ford Zodiacs had brightened the motoring scene enormously when they appeared in their vivacious party mood in the 1950s, Ford wanted to move the models more up-market. Accordingly, when Ford laid down plans for the MkIII version, that would succeed them in 1962, radical changes were to become evident. Gone were the dazzling two-tone colour schemes of the earlier Zodiacs. Now, it was believed that the new models should primarily rely on a quiet dignity, an air of reserved elegance, which it was felt would appeal more to the discerning market. As we have discussed, in 1959 the Ford range consisted of the base model, the 103E Popular (soon to be replaced by a Popular 100E), just as the 100E Anglia and Prefect were to be succeeded by the new Anglia 105E later in the year; however, this left a rather sizeable gap between the 2-door 997cc engined Ford Anglia 105E and the Ford Consul 4-door saloon with its 1703cc engine.

A brace of Zephyr 4's.
Pictured above, a late model Zephyr 4, recently re-registered with a 'J plate' is seen in the company of an ex-GPO Morris 1000 van and a Mk. I Escort at Birkby in the early 1970s. The white stripe and painted bumper are completely unauthentic. Below, a now preserved Mk. III is seen off Blackwell Road outside Carlisle. Registered in Cumbria, the car shows its rear aspect to good effect. The stylish wings and light clusters were a distinctive feature in the rear view, and the rear registration plate was hinged to conceal the fuel filler cap.
Photographs: Both Robert Berry

This gap was to be filled by an entirely new model, the Ford Consul Classic, which was sold in either 2- or 4-door saloon guise. Having already made plans for the Classic, it was felt possible to move the whole of the new MkIII model range of large Fords more up-market. Yet, at the same time they decided to retain, with suitable modifications, the drive-trains of the earlier models. There were still to be three models in the big car line-up, but the Consul name was considered to be a little low-key for the new heights which Ford hoped to aspire to with these more prestigious large cars. As a consequence, this name was cascaded from the large car range down to the family saloon market which was to be launched with the soon to be announced Classic. In view of the fact that the name Consul had long been associated with Ford's larger offerings, this obviously donated a little extra panache to the new family saloons (and undoubtedly influenced a few sales as a result)! Meanwhile, a modification of the over-head valve 1703cc in-line 4-cylinder unit that had powered the MkII Consul was used in one of the new large models that were being introduced; this car was the Zephyr 4. The former 6-cylinder engines of the Zephyr MkII and the Zodiac were also modified, and these were fitted into the new models which became known as the Zephyr 6 and the Zodiac MkIII. Actually, the new Zodiac was to have its own individuality as an executive model rather than be a luxury version of the Zephyr as before. They were then designated as model codes 211E (4-cylinder cars) and 213E (6-cylinder cars).

The new models body design had been originally created by the elite Italian styling house of FRUA, but alterations to their designs were incorporated into the front end by the Canadian stylist Roy Brown, although the rear of the car remained largely to the FRUA design. The cars shared a similar design concept, but they had a unique radiator grill for each model, whilst the Zodiac had a different styling in the rear passenger accommodation area. The styling of the cars was both totally different from earlier models and something of a compromise between Italian and American fashions. These styling considerations resulted in a long, low, wide block-shaped car with an airy cabin of full width screens. It was, in effect, a merging of lines from the prominent splayed rear wings, from where the eye could discern a styling line straight along the base of the side windows which ended at the leading outer edge of the large bonnet. From the point of the subtle rounded corner on the front wing, the body moulding returned back to be absorbed into the rear wing. The bonnet and roof panel were rather flat and separated by a wide, nicely raked windscreen of double curvature between the two slender pillars. With the curved side windows, these pillars were a major consideration which gave greater interior space and a very light airy interior that in turn offered excellent visibility. The rear doors on both Zephyr models incorporated a quarter-light behind the side window, and the rear roof pillar was noticeably wide. Meanwhile, on the Zodiac models, the rear door window was rather square and a small, triangular-shaped (non-opening) quarter light was a feature of the bodywork; whilst a slender rear roof pillar provided the Zodiac with a different rake and a little more glass area to its rear screen. From an external rear view the Zodiac was different to the Zephyrs in as much as it featured full width trim below its boot-lid, which comprised of four thin chrome bands. At the front of the car, each model had its own style of radiator grill; that on the Zephyr 4 being a simple oblong of thin, concave vertical bars with its extreme outer edges on a slight diagonal outward taper on their climb towards the bonnet with the headlights set outside. The Zephyr 6 had a radiator grill of a similar style but, being of full width, this was a little deeper and divided into two by a slender body coloured 'V' and had the headlights situated within the outer ends of the radiator grill. The Zodiac wore a grill of a similar concave pattern but it was of a full width style representing fine horizontal bars and illumination was in the form of dual headlights, each model having combined circular sidelight-indicator units in the lower front wing.

The interior was upholstered in a new soft vinyl, though leather was an option at extra cost. The seats were of the bench type with central pull-down armrests, but individual front seats were also available. Internal door locks were of the push-down button type and the internal door release was a lever under the door mounted armrest. Reversing these big Fords was simplicity itself and, although those angled rear fins housing the tail-light units were not to everyone's taste, they gave you something to aim with when negotiating a rearward manoeuvre between gateposts. Forward motion into a confined space was something else, and though you knew that there was a corner on your near-side front wing, precisely where was anyone's guess! A brilliant idea with these big Fords, after years of the infamous vacuum type windscreen wipes, was the introduction of a 2-speed electric wiper. They actually move the rain, 'I can see, I can see' one journalist wrote when they were tried out. What is more (and in the same review he wrote), 'with all that sheet steel forward of the windscreen, we got the sensation of piloting the Ark Royal around Scarborough's boating lake'. Underneath this vast steel sheet, which those who knew better called the bonnet, was the last variation of the 2553 straight six engine, but in four years time the big Fords were to be powered by 'V' engines. Incidentally, a new 4-speed gear-box was fitted to the MkIII models with synchromesh on all gears now, or if preferred the Borg-Warner automatic transmission. Various controls and warning lights were in front of the driver on a wide, mock-wood panel with the direction indicator and main beam warning lights, fuel and temperature gauges housed (with the speedometer) in a low, wide shroud above. On the Zodiac the various controls were seen to be arranged differently through that delicate looking steering wheel which, with its streamlined boss, had the appearance of a flying fish leaping up at the driver. No convertible models were offered now, and although an estate car version had been absent when the MkIII Zephyr and Zodiac first appeared in April 1962, by October the traditional Ford estate car manufacturer Abbott's of Farnham had redressed this deficit by producing a very attractive conversion.

Contrasting Views:
The two pictures on this page show to good effect the difference between the Zephyr 4 and the Zephyr 6. The most significant aspect is the dividing bar on the radiator of the Zephyr 6. The Zephyr 4, pictured above is seen in the former mining village of Newton Cap, County Durham, whilst the Zephyr 6 is a pre-launch offical picture. The most enduring aspect of these Ford models was their use in the BBC series Z Cars, where they gained some considerable celebrity. One wonders, did any of those 'stars' ever escape the scrap-man's cutting torch?

As our pictures on this page show, the Mk III estates had the look of a genuinely planned model rather than the appearance of a converted saloon car, but although the prototype model had used a steel tailgate the production models utilised a fibreglass one. Around the same time the well respected coach-builders Hoopers offered a really luxurious version of the Zodiac, this having extensive sound insulation, a fully retrimmed interior, new deep-pile quality carpets, a choice of seats and although retaining the same instrumentation the fascia was now veneered. Burr Walnut was used for this fascia and also on the matching door cappings. The boot was actually carpeted and the car was resprayed to the customers choice. Ford produced their own Flag-ship in the MkIII line-up in January 1965 and called the car the Zodiac Executive, basically the normal model but with every option fitted as standard equipment. The Zephyr Six proved to be a popular model for various Police forces in Britain during the 1960s both in conventional saloon car form and the estate car versions but cars intended for the Police Forces had the Zodiac engine fitted. Who can fail to remember the BBC series ‘Z Cars’ in which great actors like Stratford Johns, Frank Windsor and Brian Blessed (as Fancy Smith) all became well-known characters? Whilst the cars never achieved the longevity of some of these celebrities, they were the real stars of the show and it certainly was one of the greatest adverts that Ford ever enjoyed - and that on the BBC too! One final variation was by the Bolton based specialist coach-builders Coleman-Milne, who offered the Cardinal - a rather elegant hearse based on either of the two Zephyrs or the Zodiac. A fair amount of these were built, and after a few years a number of them started to be offered on the second-hand market, and as one newspaper advert stated ‘just one careful owner and few complaining passengers!’

CHAPTER 10
CORTINA MK1

Pictured Above:
The Cortina family circa 1964. This official photograph, besides revealing the photographer's shadow, also shows the 2-door, 4-door and estate versions of a car that became a market leader.

The year 1962 saw the launch of the Consul Cortina which, although designed to replace the troubled Consul Classic, was eventually to become both one of Ford's best selling British models and a very long-running marque. It was everything the Classic promised to be, a medium sized family car at a reasonable price, and excelling in its reliability - perhaps the only exception being in starting on cold mornings! This was a little quirk which they became famous for, and one that gave us many a short hard run as we endeavoured to launch our mutual friend Brian off to work at the Co-op. Except for this little idiosyncrasy, there were few teething problems for the front-engine/rear wheel drive car, as its power unit had been thoroughly tested in the latter models of the Classic. Considering the fact that its main competition, namely the transverse-engine/front wheel drive BMC 1100 series, was already an outstanding success it seems strange that Ford stuck to rear-wheel drive, but they did and the model soon started outselling its rivals. It was available as either a 2-door or a 4-door saloon, or as an estate and there was also a choice of engines, either 1200cc or 1500cc. The styling was relatively conventional, particularly when compared to the Classic, but this was probably intentional. The driver looked through a slightly curved windscreen, down to a low, wide radiator grill that reached almost to the sides of the car. On the bonnet there was a low, wide hump and a badge was fitted to this on its leading edge; on earlier models this proclaimed the car to be a Consul, but on later models it was changed to Cortina, was this again a desire to disassociate the car with its fore-runner?

On basic models the grill was formed from thin horizontal bars painted body colour, whilst the deluxe and super models had a chrome-mesh grill. Up to 1965 the oval side-lights located below the head-lights were in the leading edge of the wings; however, the discontinuation of the basic model in that year led to a re-styled grill which embraced the side-lights. The rear of the car had a deep, flat, and surprisingly spacious boot which was fitted with a rather flat lid. A set of tail-lights were enclosed in a circular cluster that was divided into three sections - indicator, reflector, and side/brake lights. It was with a certain trepidation that the circular rear lights with their three pronged lens divisions were fitted to the cars at all as it was feared that the aware public might associate them with the star that proclaims the arrival of the prestigious Mercedes Benz cars from Germany. As things turned out the tri-divided circle actually became identified with the Campaign for Nuclear Disarmament, so another first for Ford was subconsciously producing the first 'Politically Correct' car. The rear window was rather deep and full width giving excellent rear-vision. Chrome bumpers were fitted front and rear.

A feature on the sides of the Cortina was an arrowed concave flash, which was trimmed with a chrome border on the 1500 'super' following its introduction in 1963. Basic models had a monotone colour scheme, whilst 'super' models had a contrasting roof colour. On the other hand deluxe models could have either the 'super' finish, or have the concave flash painted to match the roof colour. In 1962 an experiment was made with a 2-door coupe version, the prototype of which was sent to Ford in America for evaluation. Appropriately known as the Saxon this stylish little car would have filled a much needed gap in the American small car market, but it was never put into full production and some have commented that this was due to Detroit's 'not invented here' attitude. The following year GT models were introduced in both the 4-door and 2-door versions; these had a standard Cortina 1500cc over-head valve engine, but were fitted with a Webber carburettor which increased the output to 83 brake-horse power. Naturally the GT models were also given additional appointments, including a luxury interior.

By February of 1963 Ford had developed an arrangement with Lotus to produce a sports car version of the Cortina, after this well-known sports car manufacturer had experimented with a number of Classic models (believed to be 12 in all). It was one of these twelve which one of your authors decided to pile-up at Leicester in 1969, what a waste. However, this unofficial trial with the Classic paved the way for marrying together the Cortina body-shell and the Lotus 105 BHP engine. Employing the 2-door version, the Lotus Cortinas were finished in an unmistakable two-tone paint scheme. An all-over cream finish was high-lighted by a distinctive olive-green side flash, whilst the familiar green and yellow Lotus badge was positioned on the side of the rear wings. Fitted with wider wheels, the standard Lotus Cortina was good for 108 mph and returned approximately 21 mpg. A lot of the cars also sported a black mesh radiator grill, over which would usually be mounted additional

Pictured Above and Below:
Two official pictures of the early production Cortina models; and if you look twice, you will see that it is not the same car, despite the fact that the registration plate is identical on both models. The front aspect shown above is on a Super model with the chrome side trim, whilst the picture below shows the rear aspect of a GT version, as demonstrated by the 'Grand Tourer' badge on the rear wing. This picture also shows to good effect the cicular tail-light cluster referred to in the text. Another difference between the Super and GT models are the chrome wheel trims seen in the upper view, compared with the plain wheels on the GT.

driving lights. Approximately 3,301 Lotus models were made between 1963 and 1966, though this was only a small proportion of the total Cortina MkI production which lasted until the introduction of the MkII model in 1966.

The Ford Cortina had been named after the location of the 1960 winter Olympic sports in Italy at Cortina d' Ampezzo to give the new family saloon an international air. The models were certainly enthusiastically welcomed by the car-buying British public which encouraged the Ford Motor Company to offer more variants of the Cortina than any model previously manufactured by them. By the time of the Earls Court Motor Show in 1963 there were twelve models in the Cortina line-up from the basic 2-door saloon selling at £573 to the Lotus Cortina which had a price tag of £1,100. Besides the various saloon cars already mentioned, three estate cars arrived upon the scene in March 1963; these were the 1200 De-luxe, the 1500 De-luxe, and the Super.

Pictured Below:
The Cortina Estate, became one of the most poular Ford family estates thus far produced, and although they were not built in as great a number as its nearest rival, the Morris 1000 Traveller, it did set the trend for the range of successful estates that followed. This view shows a 1965 De-luxe model, with plain sides, the imitation wood-grain finish having been discarded by someone at Ford who had a conscience. Strangely, this attempt to introduce a wood finish was a relative failure, mainly due to its un-natural look. Yet, when we compare this with the wood finish applied on the Morris rival, it is strange to relate that Ford got it so wrong in this instance. As many readers will know, the 1000 'Traveller' retained its wooden sides down into the 1970s, and has today become a collectors car. One wonders (by way of contrast) how many 'woody' Cortinas have survived?

The estate cars had heavy duty rear suspension and a one piece tailgate which made it necessary to change the large circular tail-lights (one of the most distinctive styling features of the saloon), to a vertical styled lamp unit which was worn on the end of the rear wings. All the estate cars had a cabin-like extended saloon, indeed the rear most roof pillar on them was very similar to that on the saloon. The body also had the recessed styling side-flash, but on the Super Estate, Ford's styling department went wild. Perhaps someone thought the 'Americans are fond of their woodies, lets go for it,' whatever the sides of the Super Estates were finished in imitation wood spars with a pale coloured surround styled around the wheel-arches, whilst imitation wood was also fitted to the tailgate below the rear screen. Cars fitted with this styling feature were usually painted with one of the choice of Cortina colours and many had either a cream or pale grey roof. The 'wood' finish looked far from realistic, having come direct from the Forrest of Plastic Trees, just the thing for 1960s and the yacht set, but did it catch on? Well apart for an attractive Corgi car model, it was so inspiring that by the Autumn of 1964 it was back to tin sides and the good old spray-paint. A case of thank-you America, but when in Rome....!

Crayford, the convertible specialists, did not get to work on many MkI Cortina's as they were very busy with conversions on other cars and an order for '57 varieties' of the mini based Wolseley Hornet convertible. Incidentally, these minis were painted either 'Toga White or Birch Grey' and were prizes in one of the largest consumer competitions of 1966 that was organised by one of the country's biggest food manufacturers. As usual, Crayford made an excellent job of the Ford Cortina although it is thought that only a handful were converted for the British market before the MkI Cortina was succeeded by the MkII. However, besides the home market conversions, around 30 were built for use as taxis and exported to Bermuda. Crayford also went on to make convertibles of each model of the Ford Cortina from the MkI to the MkV as each model succeeded the previous one. It is rumoured that among the MkII conversions, 20 of them were of the Lotus version. Finally, as far as our time period is concerned, this was the end of the Cortina story. Yet, the car's drive-train was continued (at first) in the totally restyled MkII which appeared in 1966. As we will tell in a later book, this MkII model had a much more squared appearance, and eventually went on to employ either a 1300 or 1600 engine. Yet, this was only the second variant in a marque which continued for over a quarter of a century until the Sierra eventually replaced the Cortina Crusader the last of the breed.

Pictured Above:
What we might call Police Performance, as five consecutively registered Mk I Lotus Cortinas appear with their drivers in a line up which also features two Mk II Jaguars, what appears to be an Austin Westminster Farina, a Triumph motor-cycle and a 'scooter'. Presumably the black grilled Lotus Cortinas were considerably faster than the little 'perfume-burner'.

Chapter 11
FORD CORSAIR 1963 - 1970

Corsair, a very word that is evocative of Motion, Flight and Freedom, and there is now a very active owners club for these Fords! However, until recently the rarity of the cars had meant that they were in danger of becoming the forgotten Fords, so it is indeed fitting that we close this book with a brief consideration of the Corsair by Robert Berry. 'The very first time I saw one, I was ten years old at the time, it was at the junction of a busy road in Huddersfield. I remember it as being a beautiful car, pale green with a cream roof, and I recall being very impressed as I watched it drive away. When my father and I had walked into the town centre, we stopped and admired others in the showroom of the main dealers, Brockholes Motor Company. It was at the Motor Show in October 1963, that the Ford Corsair or to give it's original name, the Ford Consul Corsair, first appeared. It will be recalled from our discussions earlier, that the Consul name was (by this time) being used as the entry level to the big Fords; however, by 1962 the name was considered a little down-market, so the name was attached to the early Cortina's and Corsair's but was soon deleted'. The Ford Motor Company then, as now, wanted to build cars for everyone and the Corsair was aimed at customers who wanted something a little more up-market than the Cortina but did not want a car as large as the Zephyr. Originally, all the Corsairs had a 1,498cc straight 4-cylinder engine with over-head valves, the bore and stroke being 80.97 mm x 72.75 mm. As with all Fords at this time, the Macpherson strut suspension system was used at the front while the rear of the cars used semi-elliptic springs. Disc brakes of 9.5 inches were used at the front of the cars, while the rear wheels were fitted with 9 inch drum brakes.

Picture Above:
The Corsair was another of the cars designed for the mid-sixties market, and it built on the success of earlier products. Despite its critics, this car went on to overcome teething problems and became a firm favourite with both families and business users as seen with the 2000E above.
Picture Below:
The estate version went into production in 1966, but was considerably rarer than the saloon.

The styling of the Corsair, was in truth, nothing ultra-exciting, but it did have a crisp and clean style possessing a certain harmony. This was accentuated in the way that the top wing line and bonnet swept gently down to the shallow wide radiator grill with it's slightly recessed headlights; while the bottom wing line almost mirrored the curve on it's way up. Both of these curves were emphasised by a discrete chrome trim. The overall styling was rather reminiscent of the 1962 Ford Thunderbird, and it may have been slightly influenced by USA fashion, but it was much less severe; yet as some have said 'you don't need to look twice to see her daddy was a Yank'. From 1963 until 1965, the model range consisted of the Corsair 120E and the Corsair GT, both of which were available as either 2- or 4-door saloon cars. The GT models were fitted with a high performance version of the 1,498cc engine, in which the principle changes were a four-branch aluminium inlet manifold fitted with a dual-barrel Weber carburettor as opposed to the single-barrel Zenith of the standard engine; they also had larger exhaust valves and reshaped combustion chambers. In 1965 Ford switched to V4 engines to power the Corsair's, and after this customers had a choice of either 1,700cc or a 2-litre engine; as a matter of interest the Corsair's with the V4 engine had bigger front disc brakes too.

During 1966, our old friends Abbot's of Farnham, now turned their attention to estate car conversions of the Corsair with the full approval of the Ford Motor Company. Around this time that other conversion specialist company, Crayford, were busy making convertibles from the saloon but it is believed that only around 100 were so converted. To capture every sector of the British automobile market, Ford had an executive car in every model range during the late 1960s; for example the Escort 1300E, and the MkII Cortina 1600E, so it came as no great surprise when in January 1967 the Corsair range received it's own flagship in the uniform of the Corsair 2000E. Whilst outside the general scope of this book's time-scale we might still mention that this variant was quite lavishly appointed with it's engine and luggage compartment lights, deeply upholstered reclining bucket seats, Nylon deep-pile carpeting, built-in push button radio, walnut fascia and fully carpeted boot. Obvious external differences were the black vinyl roof, fancy road wheel covers and a new radiator grill of thin horizontal bars. As with other Corsair's, Girling disc and drum brakes were fitted, the 2-litre engine with it's Weber down-draught Carburettor could accelerate the car from 0-60 in 12.5 seconds. Bodywork was usually painted in one of the shades of metallic paint that were becoming increasingly popular, giving the 2000E an even more up-market appeal. In all its various forms, the production model Corsair sold rather well through it's years, but it was quietly withdrawn from the Ford range when the Cortina MkIII was introduced. This new, larger Cortina, offered models which had been offered in the Cortina MkII and the Corsair ranges, but from now on economies of manufacturing could be more readily achieved with just one basic body pressing. There you have the case for the defence, as with all cars they have not escaped criticism, the top of the front wings would rust through on badly neglected cars, and lets face it, the early V4 engine was not without criticism. Yet, on the whole, the Corsair was quite a good car, no scorcher but then again, it was never intended to be. It was just one of the reliable family saloons which Ford had persisted in offering to the public, and based on overall levels of research, design and mass production, followed by sales and service and customer support from a network of dealers which other manufacturers would have been proud to offer!